I0815362

Chasing the MOTHMAN

Anna Anderhagen

Big Buddy Books
An Imprint of Abdo Publishing
abdobooks.com

abdobooks.com

Printed in the United States of America, North Mankato, Minnesota
102023
012024

Design: Denise Hamernik, Mighty Media, Inc.
Production: Mighty Media, Inc.
Editor: Liz Salzmann
Cover Photograph: Worldillustrator/Adobe Stock
Interior Photographs: Chase D'Animulls/Adobe Stock, p. 23; Chris Dorst/AP Images, p. 9; Dave/Adobe Stock, p. 29; Esteban De Armas/Shutterstock Images, p. 15; Frameworld/Adobe Stock, p. 11; Gregory M. Davis Jr/Shutterstock Images, p. 27; The Herald-Dispatch/AP Images, p. 17; JM-MEDIA/Shutterstock Images, p. 5; Mal Vickers/Wikimedia Commons, p. 25; OZinOH/Flickr, p. 13; 606 Vision/Adobe Stock, p. 19; thanawong/Adobe Stock, p. 7
Design Elements: Joko/Adobe Stock (tape and paper); Melica/Adobe Stock (polaroid frame); Net Vector/Shutterstock Images (map); Oleg Iatsun/Shutterstock Images (compass); Piman Khrutmuang/Adobe Stock (paper); STILLFX/Shutterstock Images (background texture); sudowoodo/Adobe Stock (silhouette); Wandeaw/Shutterstock Images (background texture); Yevhenii/Adobe Stock (tape)

Library of Congress Control Number: 2023939281

Publisher's Cataloging-in-Publication Data
Names: Anderhagen, Anna, author.
Title: Chasing the mothman / by Anna Anderhagen
Description: Minneapolis, Minnesota : Abdo Publishing, 2024 | Series: Chasing cryptids | Includes online resources and index.
Identifiers: ISBN 9781098291938 (lib. bdg.) | ISBN 9781098278830 (ebook)
Subjects: LCSH: Mothman--Juvenile literature. | Monsters--Juvenile literature. | Animals, Mythical--Juvenile literature. | Folklore--Juvenile literature. | Cryptozoology--Juvenile literature.
Classification: DDC 001.944--dc23

CONTENTS

Chasing Cryptids ... 4

What Is a Cryptid? ... 6

First Sightings ... 8

A Big, Gray Figure ... 12

Driving Home ... 14

Bridge Collapse ... 16

New Sightings ... 18

Mothman Sightings ... 20

Owls and the Mothman ... 22

An Experiment ... 24

Cryptid Keeper: Jeff Wamsley ... 26

Does the Mothman Exist? ... 28

- Glossary ... 30
- Online Resources ... 31
- Index ... 32

CHAPTER 1

CHASING CRYPTIDS

You and your friend are hiking in the woods at night. Suddenly, you hear a screech above your head. Two huge, red eyes glow in the darkness. You see a creature that looks like a man with wings.

The creature hops from tree to tree. Then it flies straight up like a helicopter! Could this be the Mothman cryptid you have been looking for?

Most Mothman sightings have occurred in West Virginia.

CHAPTER 2

WHAT IS A CRYPTID?

A cryptid is an animal that has not been proven to exist. There are stories about many different cryptids. One cryptid is called the Mothman.

Many people think they have seen the Mothman, but no one has been able to prove it. Most people **describe** the Mothman as taller than a man. It has **muscular** legs, large wings, and glowing red eyes.

Some people claim the Mothman is from outer space!

CHAPTER 3

FIRST SIGHTINGS

The Mothman was first spotted in November 1966, near Point Pleasant, West Virginia. Several men were digging a grave in a **cemetery**. One of the men saw a humanlike animal fly from tree to tree. It had red eyes and glided through the trees for more than a minute.

Point Pleasant began hosting an annual Mothman Festival in 2002. The festival features a Mothman costume contest!

Three days later, two couples took an evening drive together near Point Pleasant. They saw a tall figure standing near the road. It had fiery red eyes that glowed in the car's headlights. The creature flew after them back to Point Pleasant!

FAST FACT

The couples drove 100 miles per hour (161 kmh) back to town. The Mothman kept up with them!

The Mothman can fly quickly. But it is said to be a clumsy runner.

CHAPTER 4

A BIG, GRAY FIGURE

Also in November 1966, Marcella Bennett visited friends near Point Pleasant. When she went out to her car, she heard a strange rustling sound. A big, gray figure rose slowly from the ground. It had huge, red, glowing eyes. As it **unfurled** its wings, Bennett ran back into the house. The creature followed her to the porch. It stared in through the windows!

Point Pleasant has a Mothman statue. It was sculpted in 2003.

CHAPTER 5

DRIVING HOME

On November 27, 1966, Connie Jo Carpenter was driving home from church in Point Pleasant. She saw a tall, gray, **humanoid** creature. The creature flew directly at Carpenter's windshield before disappearing! Carpenter raced home and locked herself in her bedroom.

FAST FACT

The Mothman's wingspan is said to be 10 feet (3 m)!

CRYPTID PROFILE

NAME: Mothman

CLASSIFICATION: humanoid

COUNTRY: United States

HABITATS: forests, near water

DESCRIPTION:

— large wings
— muscular legs
— glowing red eyes
— can fly
— taller than a human man

PROVEN TO EXIST: not yet

CHAPTER 6

BRIDGE COLLAPSE

Many more people claimed to see the Mothman near Point Pleasant in 1966 and 1967. Then a **tragedy** happened on December 15, 1967. The bridge connecting Point Pleasant to Gallipolis, Ohio, **collapsed**. Forty-six people died. The Mothman was rarely seen in Point Pleasant after that.

The bridge collapsed because of a structural problem. But some people wondered if the Mothman sightings were connected.

CHAPTER 7

NEW SIGHTINGS

Since 2011, there have been hundreds of sightings of a manlike winged creature in Chicago, Illinois. Most of the sightings have been near water. One night in 2017, some friends were watching the moon. They heard a loud screech. Then they saw a gray creature flying low across a nearby harbor. It looked part human and part bat!

Several Mothman sightings have been reported on or near Chicago's Northerly Island.

MOTHMAN SIGHTINGS

ILLINOIS

CHICAGO

WEST VIRGINIA

SALEM

POINT PLEASANT

UNITED STATES

N

W

E

S

Most Mothman sightings have been in Point Pleasant and Salem, West Virginia, and in Chicago, Illinois.

CHAPTER 8

OWLS AND THE MOTHMAN

Many scientists believe the Mothman is actually an owl. Most people say the Mothman has eyes that turn red when a light is pointed at them. This is called eyeshine. **Nocturnal** animals such as cats, alligators, and owls can exhibit eyeshine. There are many owls in Point Pleasant and Chicago.

The great horned owl is one of the biggest owls in North America. Its wingspan can reach 4.6 feet (1.4 m).

CHAPTER 9

AN EXPERIMENT

Joe Nickell did an experiment to show that people could be wrong about the Mothman's size. Nickell drove **volunteers** down a dark road lined with wooden figures. He asked the volunteers to guess the heights of the figures. No one guessed correctly.

This proved that it is hard to judge size at night. So, the Mothman could be something smaller, such as an owl.

Joe Nickell is a paranormal investigator. He works to uncover the truth about cryptid sightings.

CHAPTER 10

CRYPTID KEEPER: JEFF WAMSLEY

Jeff Wamsley lives in Point Pleasant. He is the **curator** of the Mothman Museum. He collects items about the Mothman. These include books, **sketches**, police reports, and letters. Wamsley believes all the witnesses in Point Pleasant saw something. He just doesn't know if it was a bird or the Mothman.

The Mothman Museum features props from a movie about the Mothman.

CHAPTER 11

DOES THE MOTHMAN EXIST?

Hundreds of people think they have seen the Mothman. There are many **sketches** and a few photos. But no one can prove that the Mothman is real.

Most scientists don't think the Mothman exists. If it did, there would be more **evidence**. These scientists believe people are mistaking owls for the Mothman. What do you think?

What would you do if you encountered the Mothman at night?

GLOSSARY

cemetery—a place where dead people or pets are buried.

collapse—to fall down suddenly.

curator—a person in charge of a museum.

describe—to tell about something with words. Such a telling is a description.

evidence—facts that prove something is true.

humanoid—having human form or characteristics.

muscular—having strong, well-developed muscles. Muscles are body tissues, or layers of cells, that help the body move.

nocturnal—active at night.

sketch—a rough drawing.

tragedy—a sad and shocking event.

unfurl—to unfold or open.

volunteer—someone who helps others in their own free time without pay.

ONLINE RESOURCES

To learn more about the Mothman, please visit **abdobooklinks.com** or scan this QR code. These links are routinely monitored and updated to provide the most current information available.

bats, 18
Bennett, Marcella, 12
bridge collapse, 16

Carpenter, Connie Jo, 14
cemetery, 8

evidence, 26, 28
experiment, 24

features, 4, 6, 8, 10, 12, 14, 15, 18

habitat, 15

Illinois, 18, 19, 20, 21, 22

map, 20, 21
Mothman Festival, 9
Mothman Museum, 26, 27
movement, 8, 10, 11, 12, 18

Nickell, Joe, 24, 25
North America, 23
Northerly Island, 19

Ohio, 16
outer space, 7
owls, 22, 23, 24, 28

Point Pleasant, WV, 8, 9, 10, 12, 13, 14, 16, 20, 21, 22, 26

scientists, 22, 28
similar animals, 18, 22, 23, 24, 26, 28

United States, 15, 20

Wamsley, Jeff, 26